ANOTHER BOOK

AFTER ED-WERD REW-SHAY

ZOË SADOKIERSKI

2015

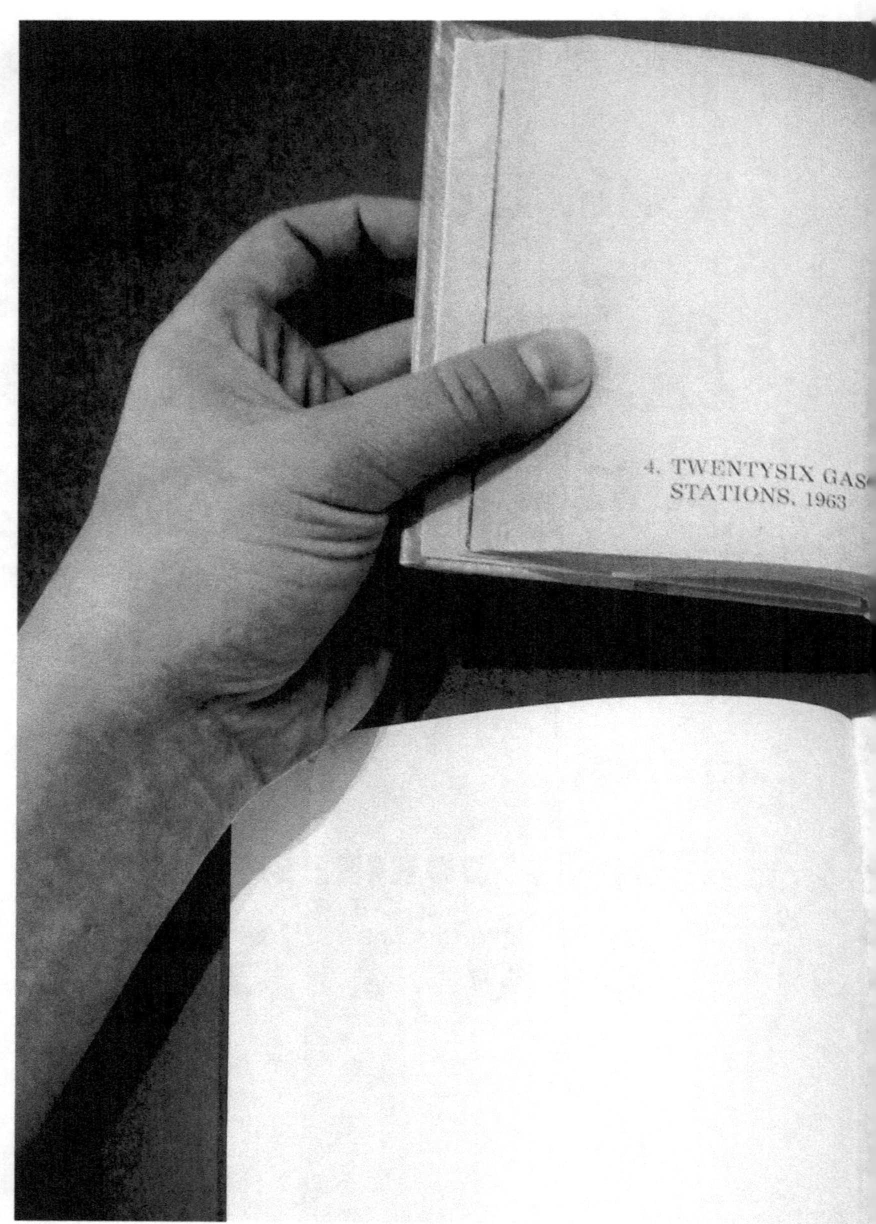

4. TWENTYSIX GAS
STATIONS, 1963

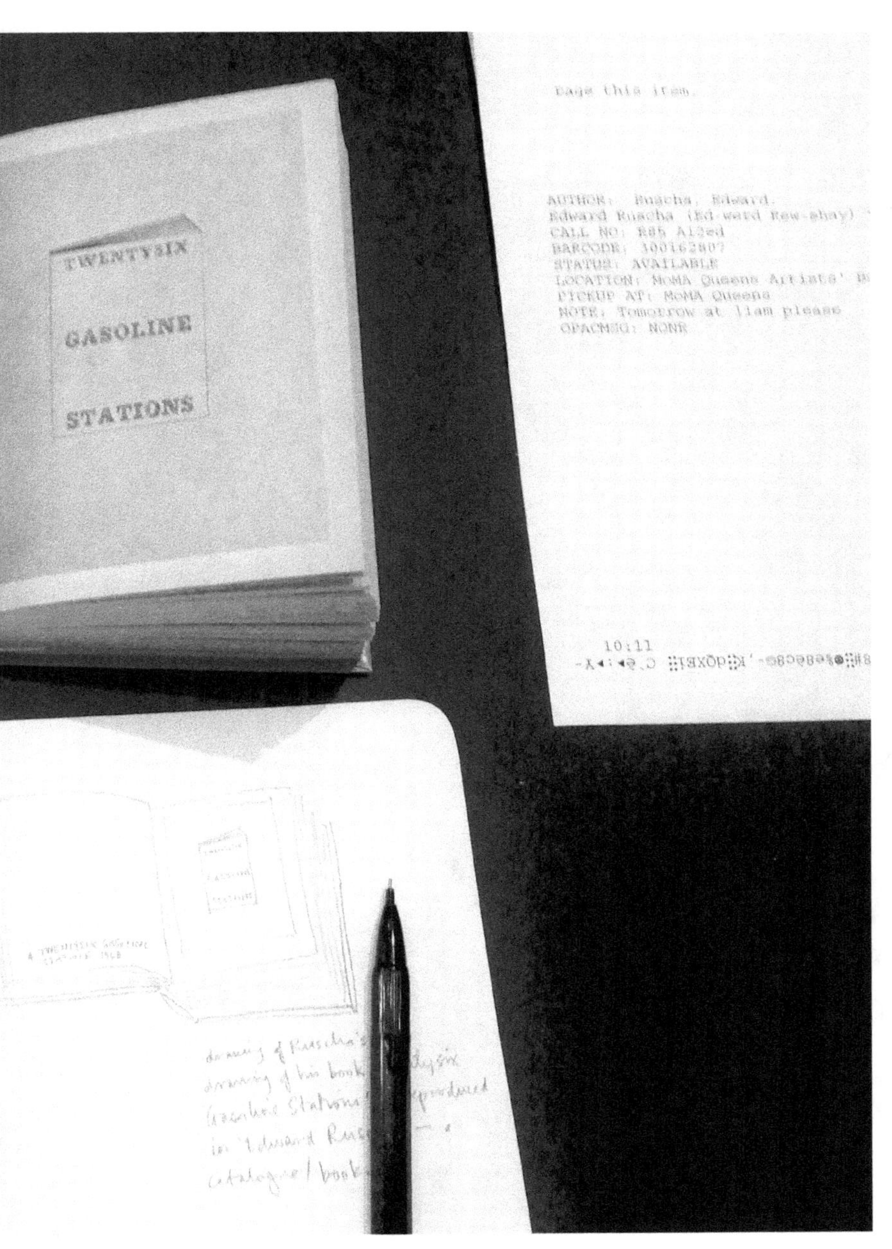
TWENTYSIX
GASOLINE
STATIONS
bags this item.
AUTHOR: Ruscha, Edward.
Edward Ruscha (Ed-ward Rew-shay)
CALL NO: R85 A12ed
BARCODE: 100162807
STATUS: AVAILABLE
LOCATION: MoMA Queens Artists' B
PICKUP AT: MoMA Queens
NOTE: Tomorrow at 11am please
OPACMSG: NONE
11:11
10:11

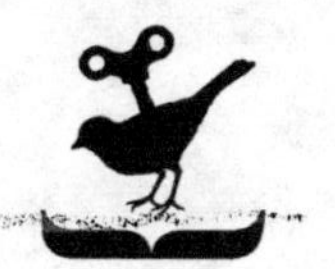

Created as part of a residency undertaken at the
University of Technology Sydney Library in 2015.
See: www.cargocollective/thebookis

Set in Stymie; after Ruscha. Except the introduction.
Sorry Ed, couldn't do it. That's in National, it's better
for longform reading.

First edition of 5 (with two proofs) produced on McNally
Jackson's Espresso Book Machine in New York City.
This edition produced using print-on-demand platform
Ingram Spark.

This book was always meant to exist, but it wasn't supposed to be about Edward Ruscha. To explain how the book came to be, I need to back up. Make yourself a cup of tea, this may take a while.

BOOK MACHINES

I came to New York for five weeks over March/April 2015 for two reasons. Primarily to research the history of artists' books and independent (small press) publishing. I've been producing artist's books for more than a decade, though in order to discuss my bookwork in an academic context — which my job as a university lecturer requires me to do — I need a scholarly understanding of the field. New York, with a diverse independent publishing culture and a wealth of research libraries, is an ideal place to develop this understanding.

The second reason I came to New York was to produce a book on McNally Jackson's Espresso Book Machine. During a visit in 2013 I discovered that McNally Jackson has a clever little machine, between the magazine section of the bookstore and their café, that can print, collate, cover and bind a book in a matter of minutes. In theory, in the time it takes to have a cup of coffee. Magical.

There are Espresso Book Machines all over the world according the manufacturers website: *Canada (11); Dominican Republic (1); Egypt (3); France (6); Japan (2); Netherlands (2); Philippines (1); UAE (2); USA (38).* [1] Disappointingly, none in Australia. So I vowed to return to New York and make a book. Here I am.

For the past few years I've been experimenting with ways to independently publish books using different print-on-demand platforms. [2] Print-on-demand is a digital printing service that allows anyone with a computer and credit card to publish a book in editions as small as a single copy. The process is simple:

the author uploads digital files of the book pages and cover to an online platform (a print-on-demand supplier such as Blurb.com or Lulu.com), then when someone orders a copy, the book is printed and posted to them, anywhere in the world.[3]

Print-on-demand means an author can publish a book without spending money other than the cost of ordering a single proof copy – as little as a few dollars. When someone orders a copy of the book online, the author receives royalties. No initial financial outlay for the print-run, no warehousing and distribution cost or management. It goes without saying that this is a game-changer for the book publishing industry.

Print-on-demand technology (and the internet) enabled me to launch my own small publishing venture, Bookwork Press.[4] Many similar small presses are mushrooming up around the world. There is a clear parallel between the current small press movement and what was happening in the art publishing world around the time Edward Ruscha was producing his 'democratic multiples' in the early 1960s and 70s – but we'll come to Ruscha shortly.

For me, the appeal of the Espresso Book Machine, in contrast to the internet print-on-demand suppliers I've previously used, is that I can watch my book being made. Using an online supplier, my book could be printed anywhere in the world and the turnaround from ordering a book to it arriving at my door is between three days and three weeks. The Espresso Book Machine promises about three minutes, and I can watch and listen to it come into being.[5]

From Sydney, I planned to document my time in New York through writing, drawing and photography, to generate content for a print-on-demand book I would produce before leaving. I had an idea for a theme or structure that would be a typology of New Yorks based on writing by E.B. White and Liz Danzico. But this idea was abandoned when I found Ruscha in the MoMA library, and everywhere else I looked.

ARCHIVAL DIGS

Back to my primary reason for being in New York – researching artists' books and the small press movement. I wanted to spend time considering ways other artists employ or exploit the book form in order to reflect on my own bookmaking practice. Although there are many excellent publications and online resources which document the field of artist's books[6], to truly appreciate an individual artist's book you need to experience it in person.

The book is a intimate form. We cradle a book in our hands, hold it close to our chest and hang our heads over it. The size and shape of the book, its weight, the paper stock, the quality of printing and design all affect our reading experience. Usually, we read alone. When we're ready, we turn the pages sequentially to move ahead.

Artists who work in book form play with these factors: within an artist's book, the format may convey more information than the written text (if there is any written text at all); paper stock may indicate how to navigate the story; the next page may reveal something entirely unexpected. To understand an artist's book, you need to experience it in its original form.

To begin my research, I arranged visits to the Museum of Modern Art library. The artists' book collection is primarily held in the MoMA QNS building, which is only open to researchers on Mondays between 11am and 5pm. The library holdings include about 300,000 books and exhibition catalogues. With limited time and limitless possibilities, I needed to start somewhere.

Johanna Drucker's seminal book *The Century of Artists' Books* cites the post-war artists Dieter Roth and Edward Ruscha as key figures in the history of artists' books. Both artists made series of projects exploring the potential of producing artworks in book form, and moreover, "Roth and Ruscha also serve conveniently to contrast the two distinct attitudes toward artists' books as an inexpensive edition or democratic multiple." (71)

I began with Dieter Roth. Drucker describes Roth as an artist working with the book as a physical form, playing with structural investigations of the book.

Roth layers pages and images of pages to create patterns and new composi-
tions within his books. In the front matter of his book *Snow*, Roth provides a
typewritten page with:

A few of the successfull recipes

offered by Rot in this Volume:

ON THE VISUAL ARTS

"Take a thing and put it on one thing
Take a thing and put it on the 2 things
Take a thing and put it on the 3 things
Take a thing and put it on the 4 things
Take a thing and put it on the 5 things
Take a thing and put it on the 6 things
Take a thing and put it on the 7 things

. . . .

sell any time"

In my studio at home, I have a folder of abandoned photocopies salvaged
from the bin at work. I kept the edges, where the pages of the book and its
cover creep into the blackness of negative photocopier space. Sometimes
there's a wrist and the heel of a hand, evidence of the person photocopying
pressing the book into the glass. I use these scraps in collages. I get Roth.

Using ephemera from the Franklin Furnace Artist File of Miscellaneous
Uncatalogued Material for Roth, I followed his instructions:

large work on paper titled 'Survey Map of South Bunker Hill Area of Los Angeles, about 1956' from the Huntington Library reminded me of Ruscha, then I turned the corner and met with another copy of *Every Building on the Sunset Strip*. At the Gagosian shop, a vitrine held an almost complete selection of Ruscha's books. A label taped to shelf Chelsea bookstore Printed Matter read 'After Ruscha'; the shelf hosted a cluster of books made in homage to Ruscha.

More than Ruscha's work itself, I am fascinated by these homage books. In 2013, MIT press published *Various Small Books: Referencing Various Small Books by Ed Ruscha*. The book collates ninety-one projects by artists who have appropriated or paid homage to Ruscha's books, and includes an appendix listing all known Ruscha book tributes. Included are projects in which artists rephotograph Ruscha's work – such as *Thirtyfour Parking Lots, Forty Years Later* (Susan Porteous 2007) – or riff off his conceptual and formal template – *Various Unbaked Cookies and Milk* (Marcella Hackbardt 2010).

Drucker mentions one of these: "The parody by Jeffrey Brouws demonstrates the cult status of Ruscha's books among artists' books." (77) Parody implies the mimicry is done for comic or satirical effect. I sense there's something more than that happening here.

To think this through, I chose two homage *Gasoline* books to look at closely in relation to Ruscha's original: Brouws' 1992 publication *Twenty-six Abandoned Gasoline Stations*, mentioned above, and Michalis Pichler's 2009 *Twentysix Gasoline Stations*.

I started by thumbnailing Ruscha's book, followed by the other two. This is a technique I use to help me analyse the way each page of a book is composed, and the way the sequence of pages contains rhythms or surprises. Usually, you thumbnail to figure out how to design a book, but I find the reverse gives a particular perspective for analysing a book. By drawing each page, I slip into a kind of design trance that helps me look closely and purposefully.

I also use these thumbnail maps to compare the books to each other.

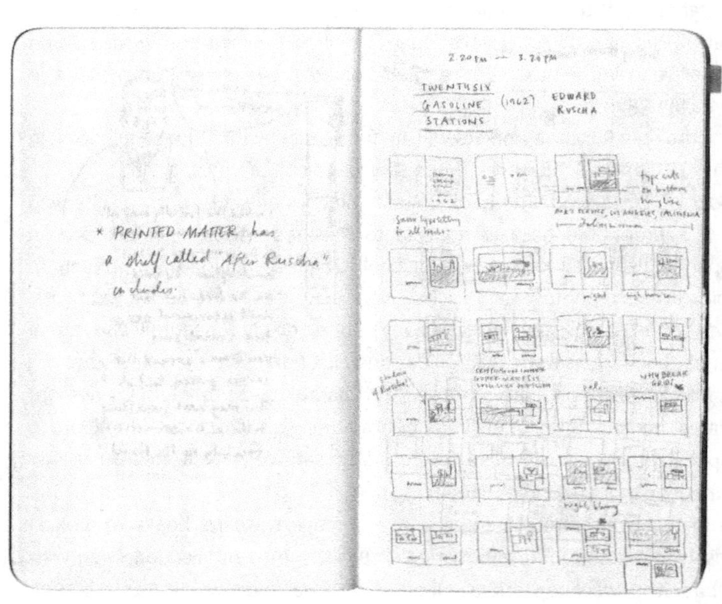

ABOVE:
Thumbnails of Ruscha's
Twenty-six Gasoline Stations
Sketched on March 23
at MoMA Library, QNS

RIGHT:
Thumbnails of Ruscha's
Twenty-six Gasoline Stations
Sketched on April 10
at MoMA Library, Midtown

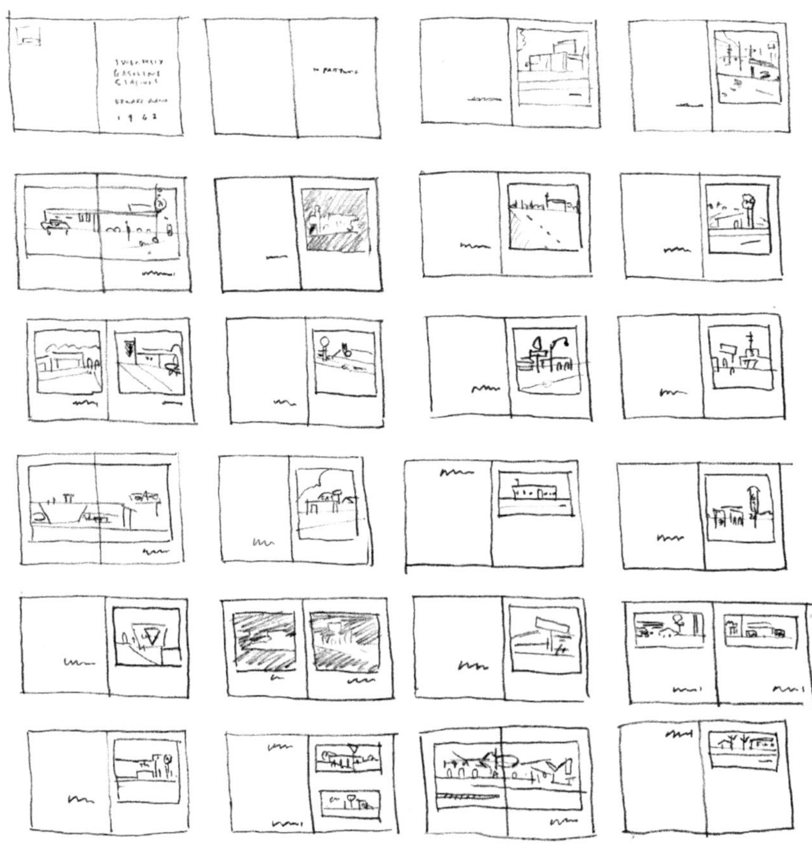

Jeffery BROUWS ~~Ruscha~~ (1992)

→ same as
Ruscha

night

MICHALIS PICHLER (2009)

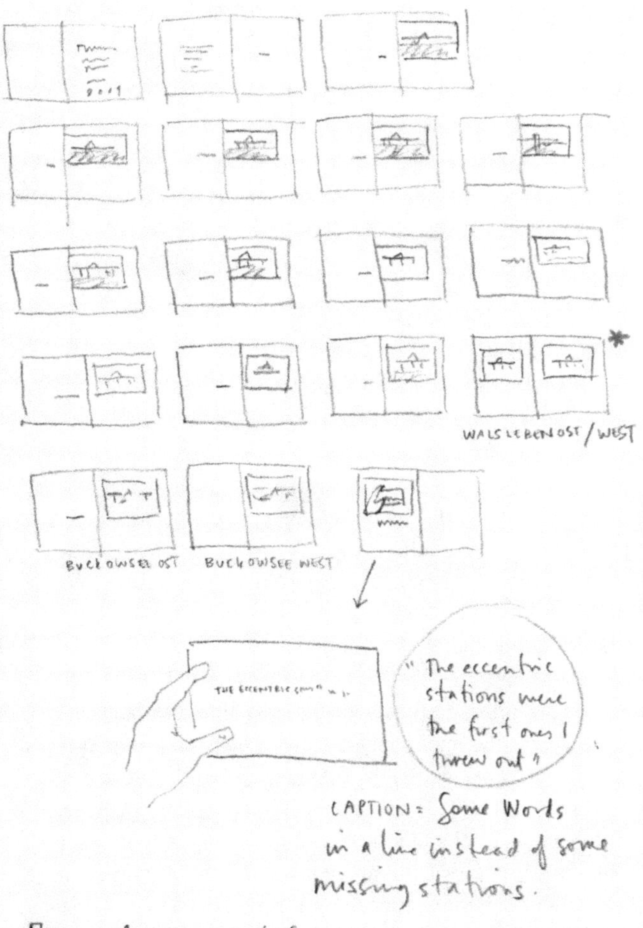

WALSLEBEN OST / WEST

BUCKOWSEE OST   BUCKOWSEE WEST

" The eccentric
stations were
the first ones I
threw out "

CAPTION = Some words
in a line instead of some
missing stations.

⇨ only 16 stations.
all the same architecture.

Brouws follows Ruscha's template exactly – using black and white photographs and mimicking the alternating square and rectangle images, and exact composition for each double page spread. The half-title page dedicates the book to Ruscha. A loose, translucent sheet inserted in the book reads:

In late twentieth century postmodern America
the number of gasoline stations across the nation has dwindled.
With rising land values and new, tougher environmental restraints,
many independent station owners found themselves unable to
replace aging underground tanks that the EPA deemed unsafe.
Subsequently, a favorite public place
where flat tires were fixed
and the weather discussed,
is slowly vanishing
from our landscape.

Brouws book is an homage to Ruscha, but it also extends the original book by overtly layering additional information. The addition of 'abandoned' to the title signals that there is something more to this book than a direct copy. The sheet of paper at the end makes this explicit. Brouws takes a familiar book and slightly subverts it, to communicate a point.

Pilchler's book follows the formal and aesthetic pattern of Ruscha's book less closely, and is less clear in its intention. The photographs are printed in colour and all depict the 'same' station in different locations – a chain of generic German gas stations that follow an architectural template. There are only sixteen stations photographed. A seventeenth photograph shows a hand holding a white card printed with the statement: "The eccentric stations were the first ones I threw out." The caption to the image reads: "Some words in a line instead of some missing stations."

I read Pilcher's book as comment on the uniformity of contemporary German gasoline stations; the stations are visually similar, and Pilcher tells us he has deliberately omitted 'eccentric' stations. I suspect there is more going on – Rushca uses cards printed with text in other books.

Where Brouws follows Ruscha's original remarkably closely and spells out his intention, Pilcher borrows the general idea and follows Ruscha's form only loosely. As Drucker compared Dieter Roth and Ed Ruscha's bookmaking to demonstrate different approaches to using the book as an art form, I compare these books as different approaches to bookwork homages.

The inevitable next step was making my own.

TWENTY-SIX VIEWS FROM THE 7-TRAIN

I have taken prosaic photographs from train windows for years. During a three-week journey on the Trans-Mongolian Rail from Beijing to St Petersburg I captured thousands blurry landscapes and desolate train platforms. Like almost everyone else with a camera and a history that involves air travel, I also have collections of airplane window snapshots: cloudscapes, sunsets cut diagonally by a plane wing, city grids.

I began this train window series on my way home from my first MoMA library session, before I thought of making this book. I was inspired, alert, overwhelmed after a day of intense study, and I took the photos as a way to channel my fidgety energy. It occurred to me later that this was the closest thing to Ruscha's theme as I was going to be able to pull off in the week between deciding to make the book and printing it.

Ruscha's *Twenty-six Gasoline Stations* follows Route 66 between Oklahoma and Los Angeles. *Twenty-six Views from the 7 Train* follows the train route between the MoMA QNS Library and my (temporary) apartment.

Ruscha's book documents gas stations; markers along a lengthy and often monotonous stretch of road. Although some shots appear to be taken at speed, for the most part the stations are stops on a familiar journey – this was a road Ruscha travelled frequently between his home town and his adopted home of LA.

My photographs capture whatever was outside when I had a clear shot of the train window, which was infrequent, especially in peak hours. Unlike Ruscha, the landscape is unfamiliar to me, even after three return trips (plus additional trips on the same train line to visit friends in Queens).

Ruscha produced his books using the cheapest, most convenient printing available to him. At the time, that was offset printing. He visited the press – the story goes that he changed the layout 50 times at the printer.

I produced a first edition of this book using the cheapest, most convenient printing option available to me – the Espresso Book Machine at McNally Jackson, on Prince Street in SOHO. There were five copies printed, with two proofs. I kept one copy for myself and gave the others to: Johanna Drucker, Steve Clay, David Senior, UTS Library, Orville Robertson and Hermann Zschiegner. This second edition has been produced using print-on-demand platform Ingram Spark, as part of my ongoing search for the ideal production and distribution model for small presses.

*

[1] The company behind the machine is On Demand Books, founded in 2008 by Jason Epstein (former Editorial Director of Random House and co-founder of the New York Review of Books), Dane Neller (former CEO of Dean & Deluca) and Thor Sigvaldson (background in advanced technology). http://www.ondemandbooks.com/ebm_locations.php

[2] In my 2014 exhibition 'Books On Demand', I displayed 12 illustrated books I created and published using different print-on-demand services (primarily Blurb and Lulu). See <http://zoesadokierski.com/Books-On-Demand-exhibition>

[3] Writing this now, I wonder if in a year or so explaining print-on-demand books will seem like explaining where milk comes from – something only wilfully naïve people will not know.

[4] Bookwork publishes visual essays, with an emphasis on integrating design and illustration (photographic, typographic, illustrative) within the primary text to create complex reading experiences. www.bookworkpress.com

[5] It should be noted that most stores have a waiting list of several days for the books to be printed, due to file checking and a back-log of orders. And although I can't imagine not wanting to watch the machine at work, others may just want their book.

[6] As a start see: *The Century of Artists' Books*, Johanna Drucker (1994/2004); *Booktrek: Selected Essays on Artists' Books since 1972*, Clive Phillpot (2013); *A Book of the Book*, eds. Steven Clay and Jerome Rothenberg (2000).

TWENTYSIX

VIEWS FROM

THE 7 TRAIN

23 March 2015. 17:02:11

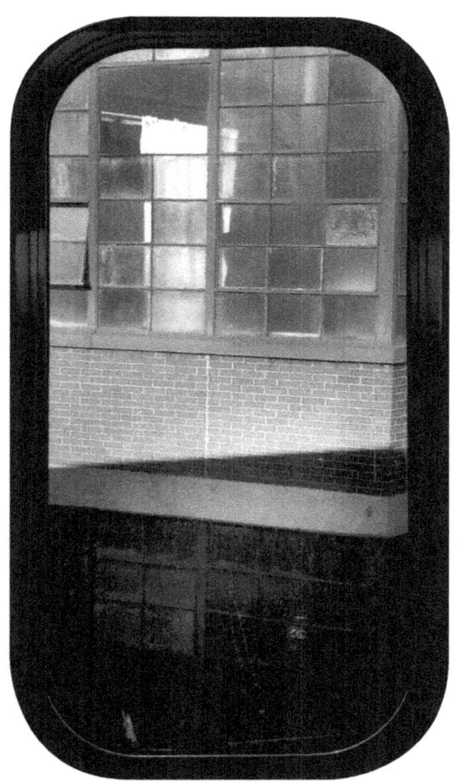

23 March 2015. 17:05:41

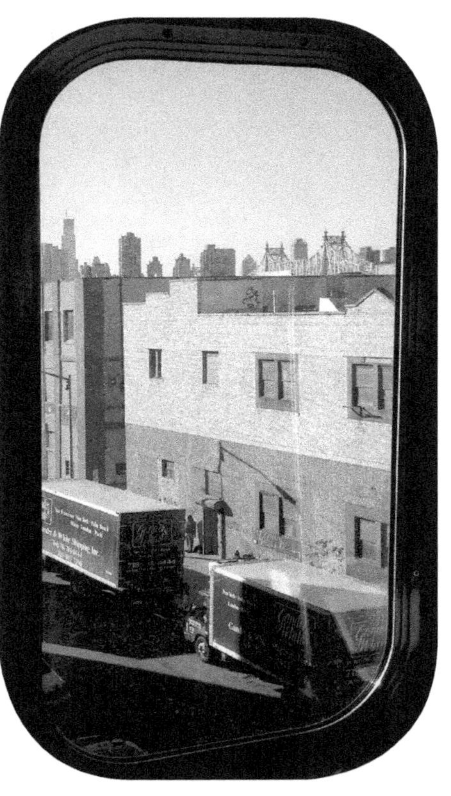

23 March 2015. 17:06:53

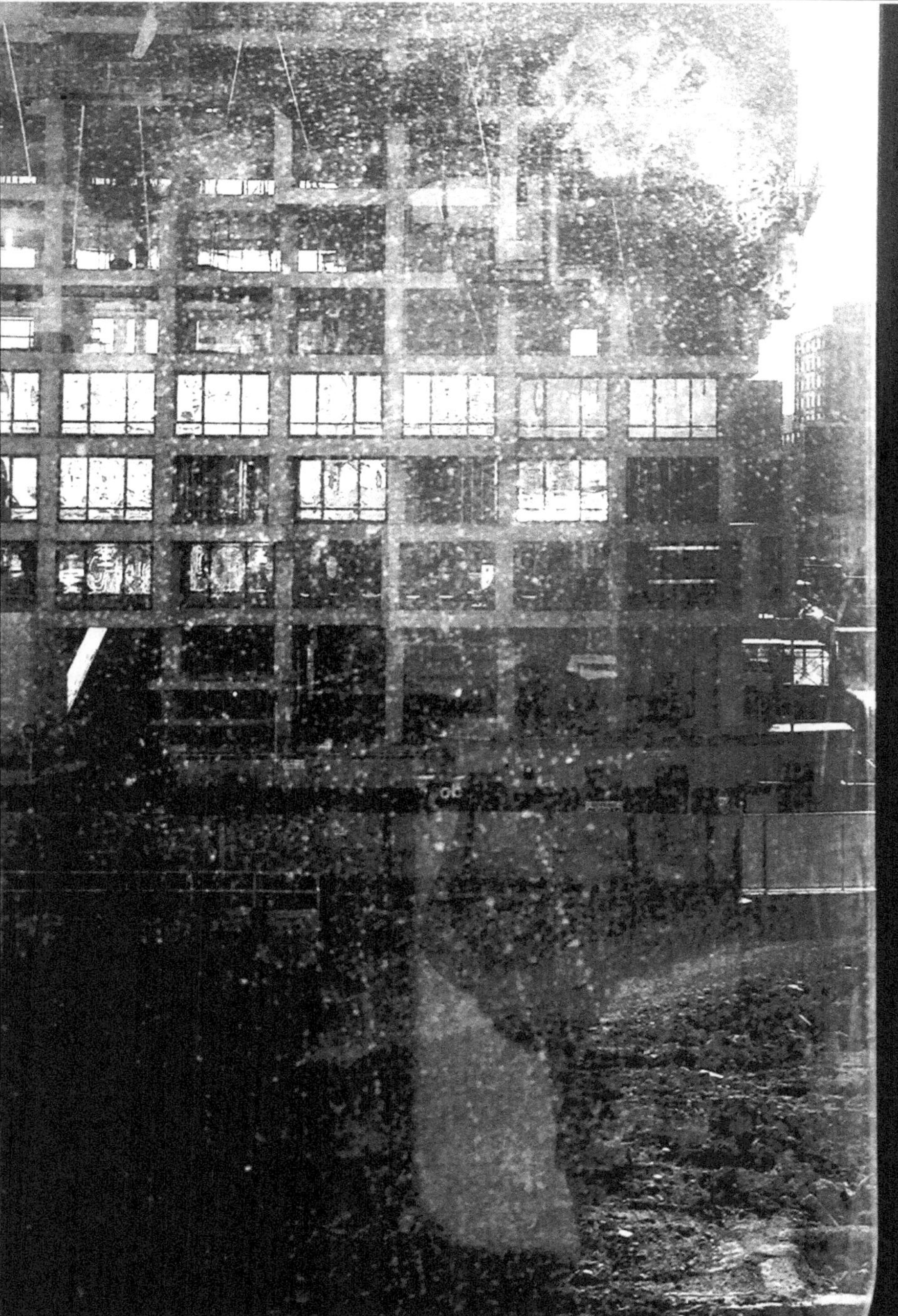

23 March 2015. 17:07:33

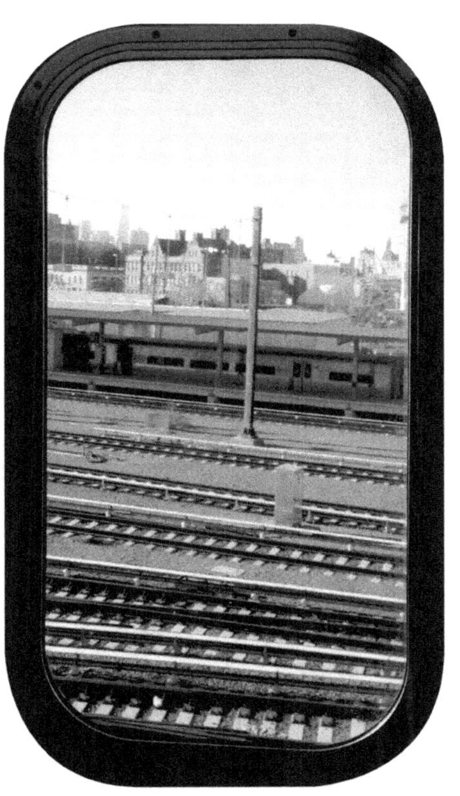

30 March 2015. 11:19:34

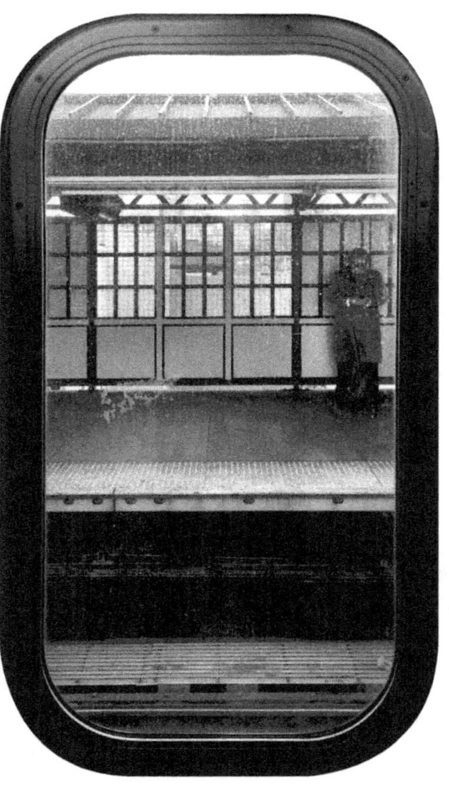

30 March 2015. 11:20:18

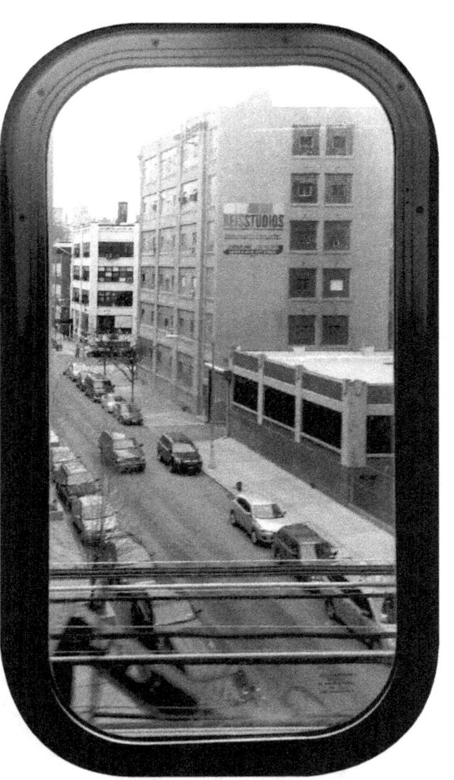
STUDIOS

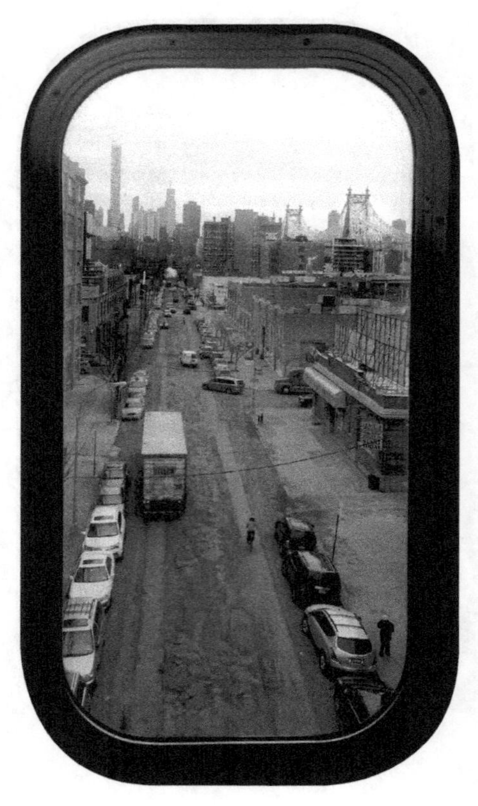

30 March 2015. 11:20:25

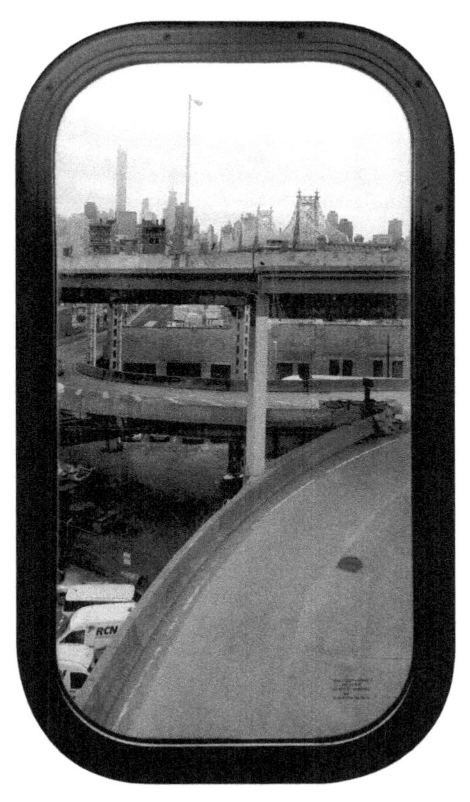

30 March 2015. 11:20:36

30 March 2015. 11:20:45

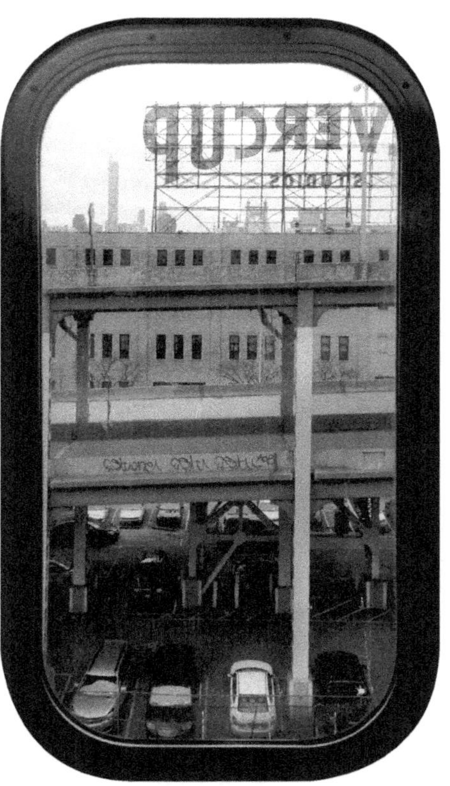

30 March 2015. 11:20:55

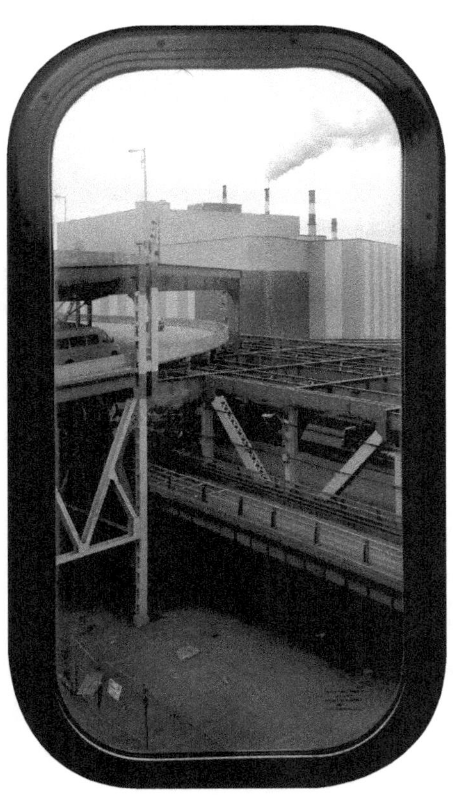

30 March 2015. 11:22:02

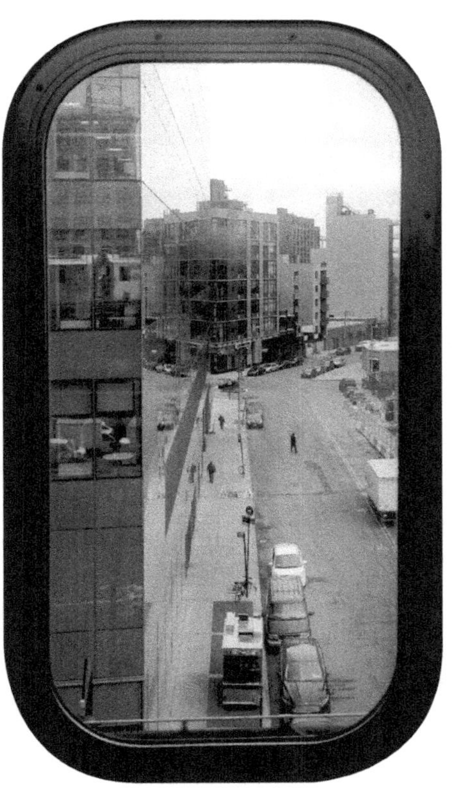

30 March 2015. 11:22:08

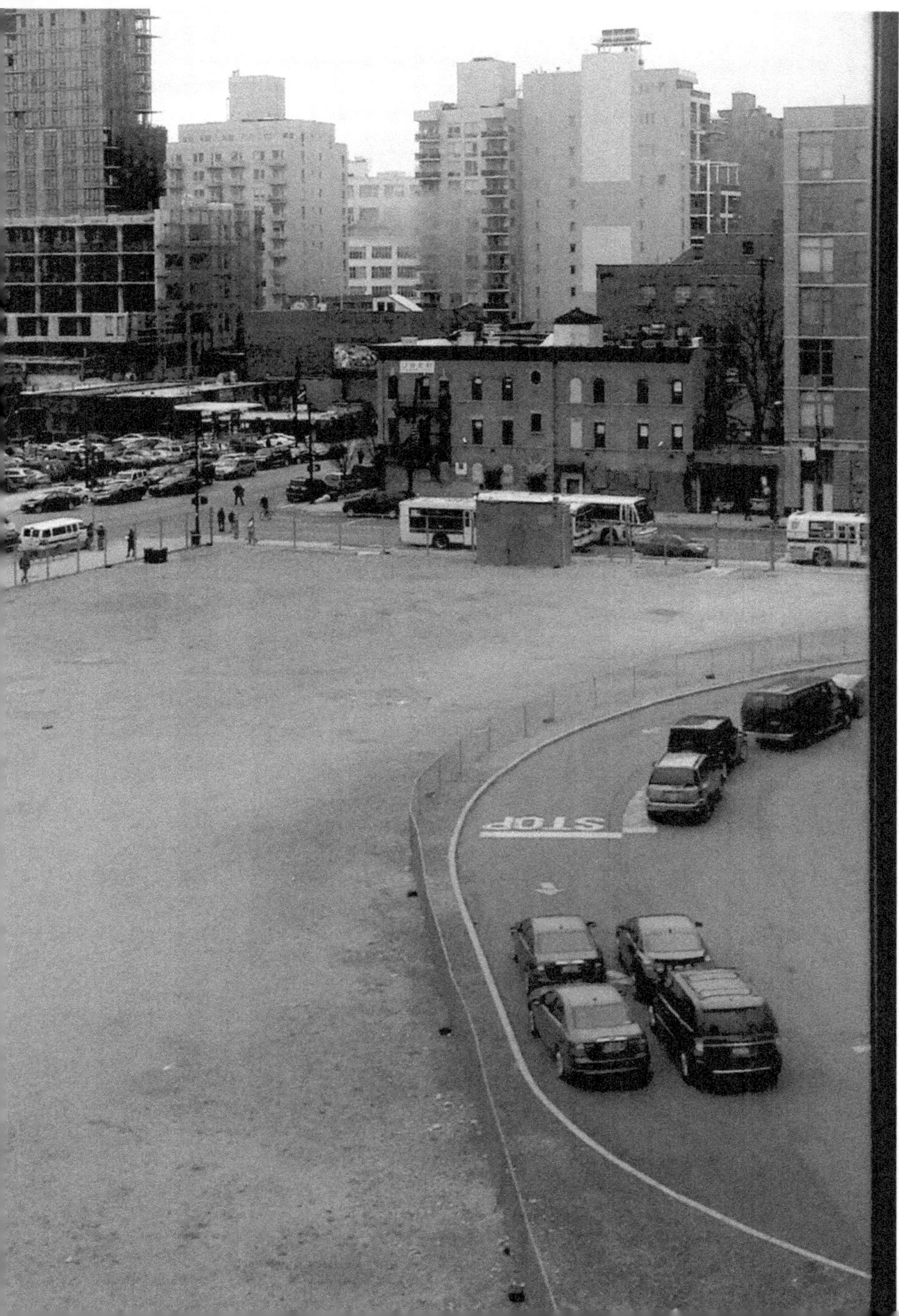

30 March 2015. 11:22:19

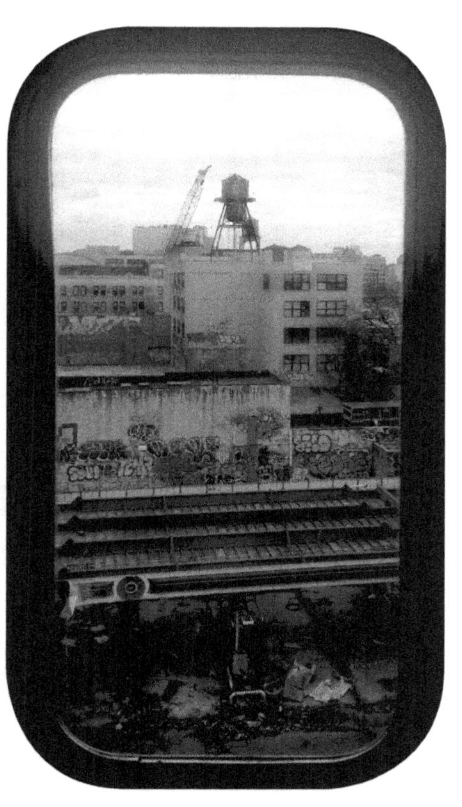

30 March 2015. 11:22:43

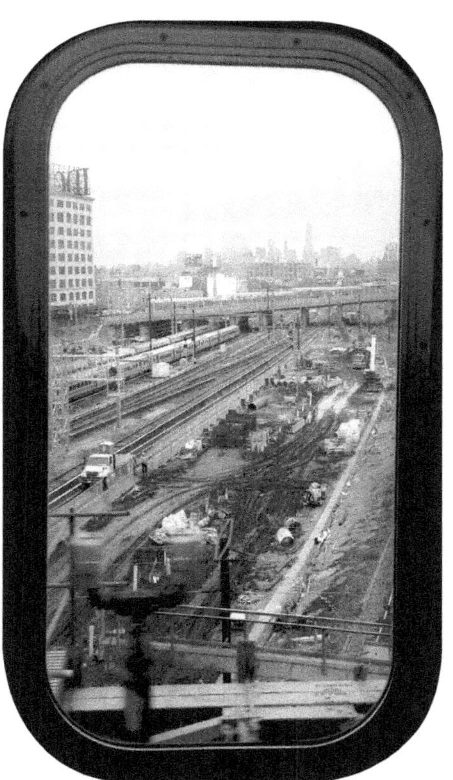

30 March 2015. 11:22:47

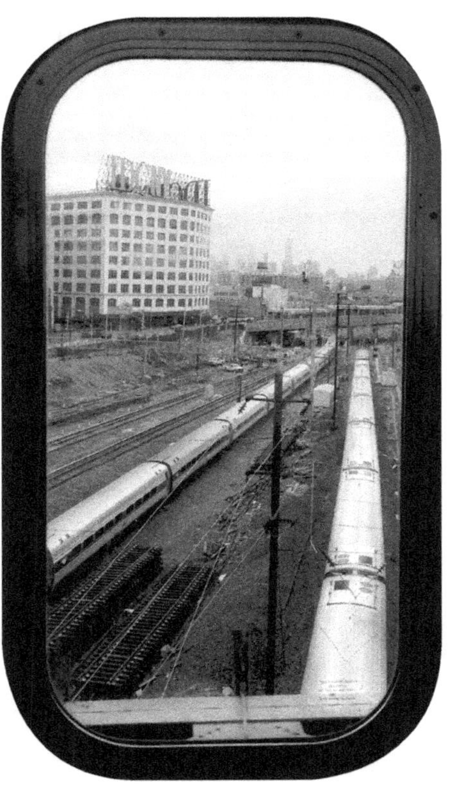

30 March 2015. 11:23:02

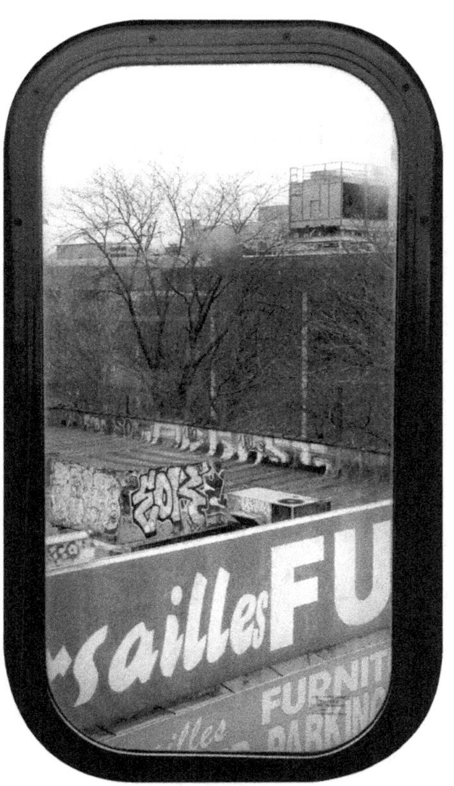
sailles FU
FURNIT
PARKIN

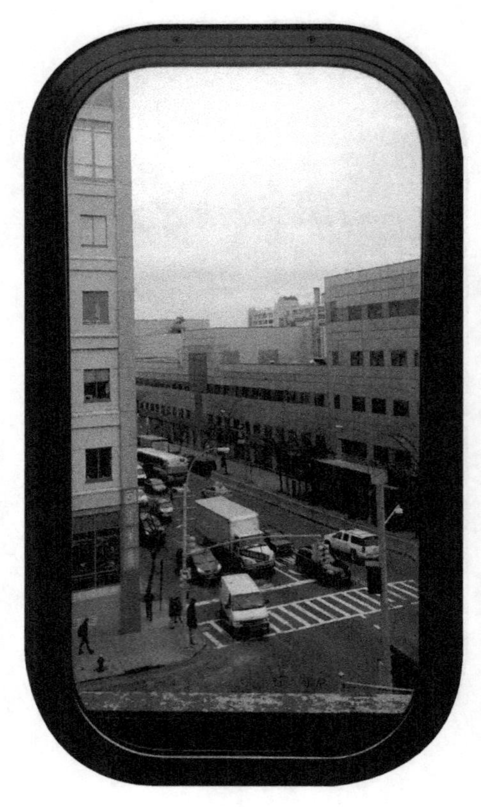

30 March 2015. 11:23:23

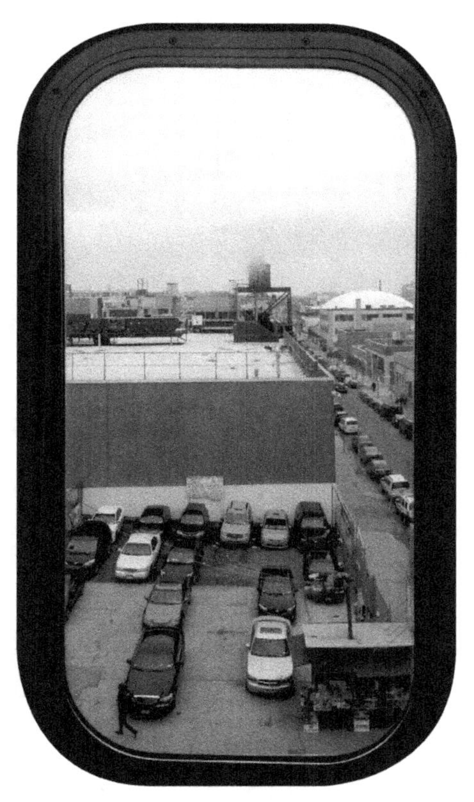

30 March 2015. 11:23:34

30 March 2015. 11:23:38

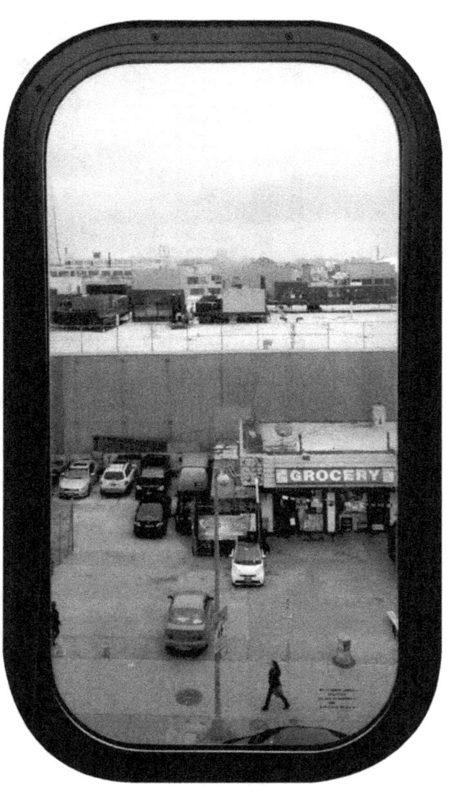
GROCERY

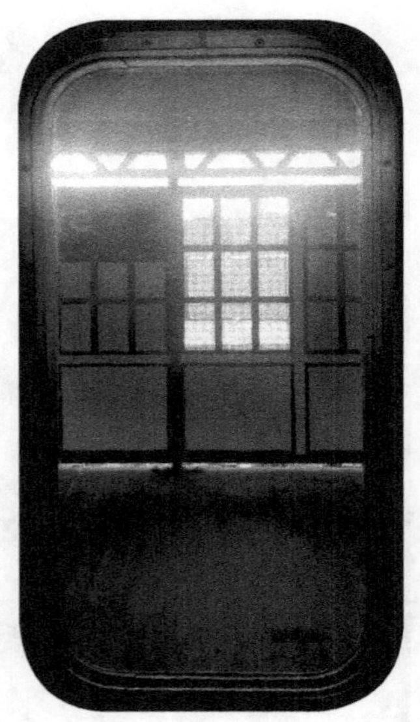

30 March 2015. 16:33:46

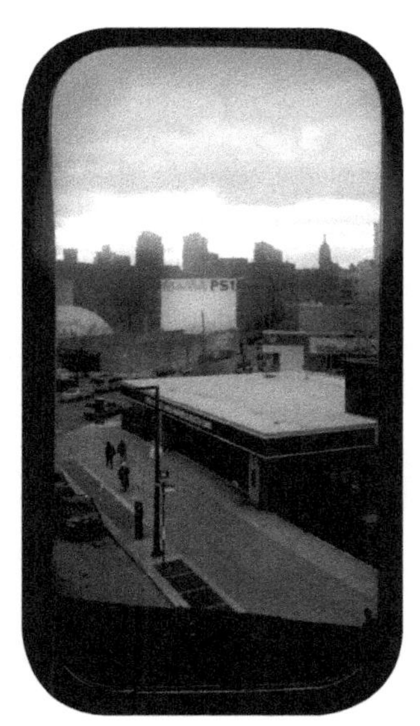

30 March 2015. 16:34:11

30 March 2015. 16:34:21

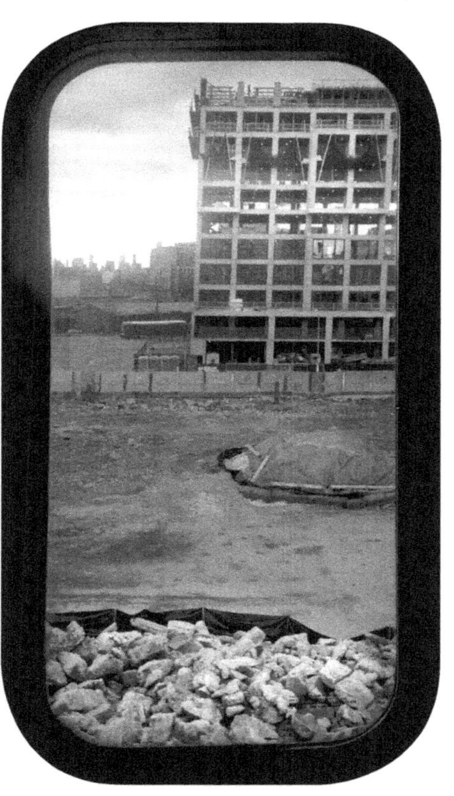

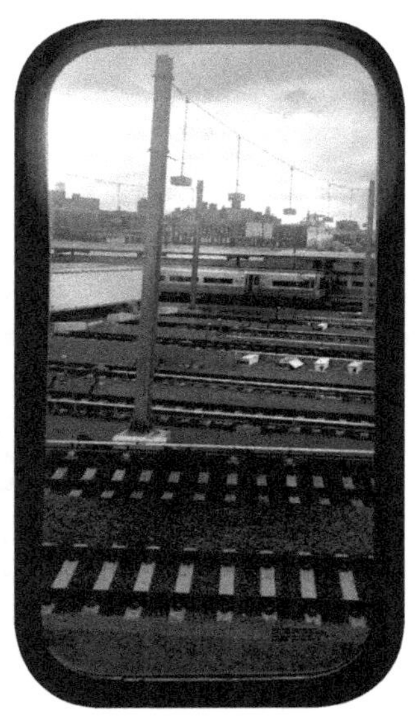

30 March 2015. 16:34:35

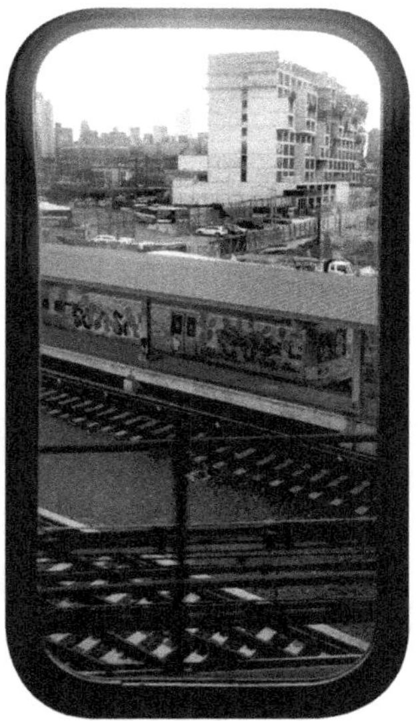

30 March 2015. 16:34:40

30 March 2015. 16:34:40

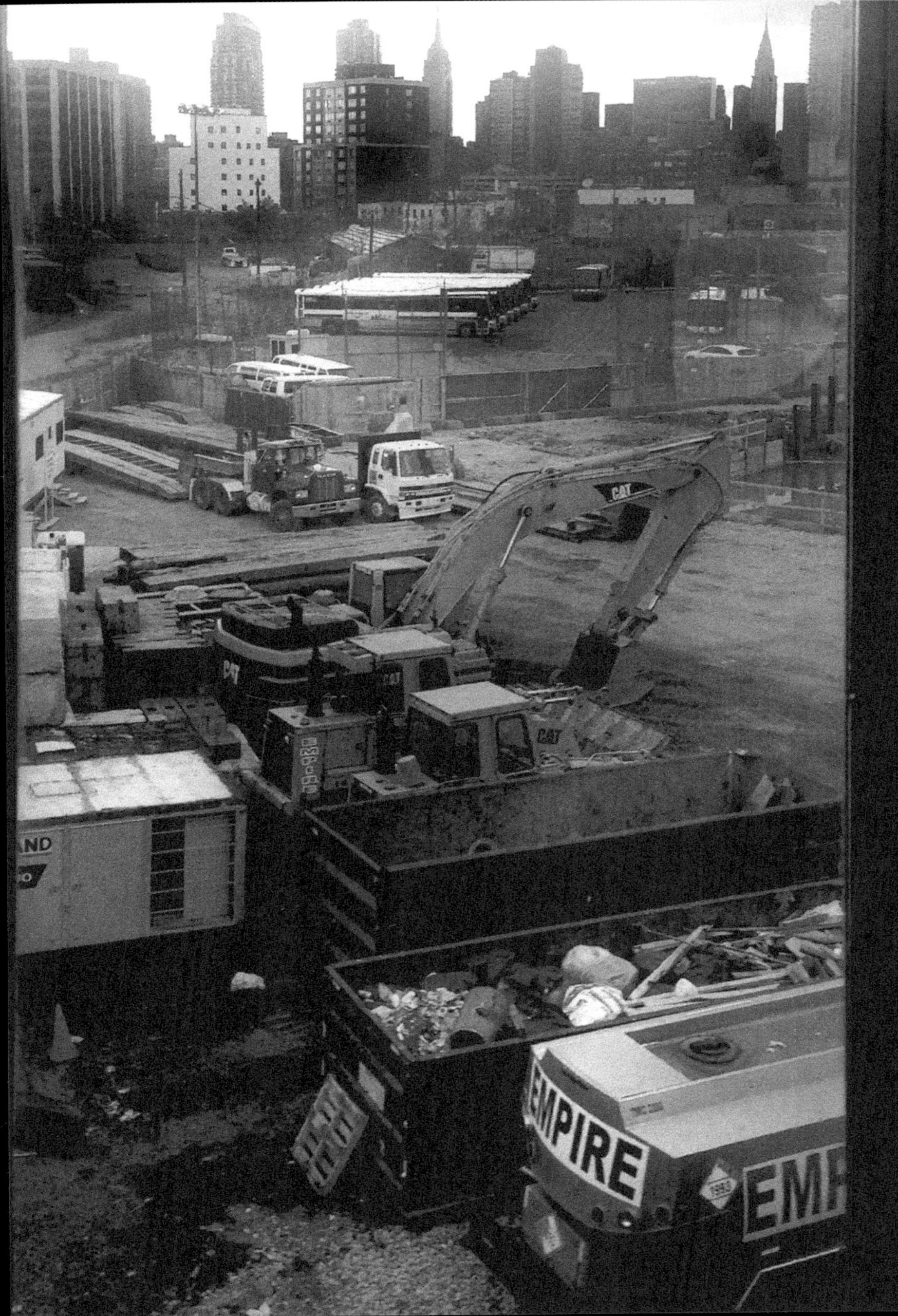
CAT
CAT
CAT
AND
EMPIRE
EMPIRE
EMP

30 March 2015. 16:35:04

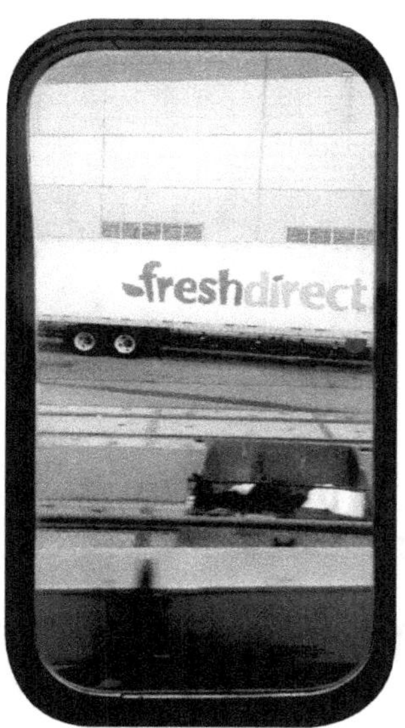

freshdirect

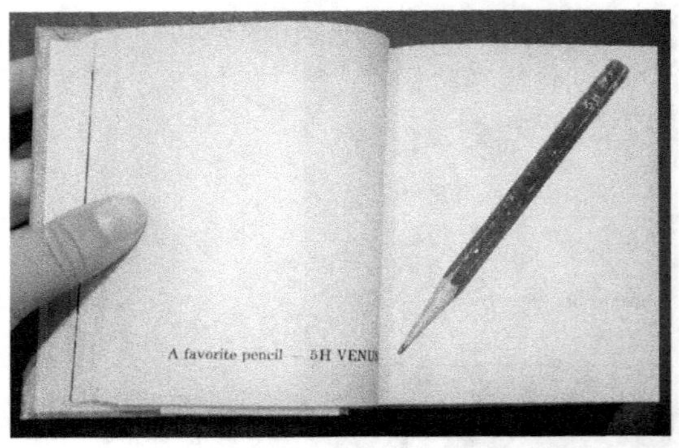

ABOVE: RIGHT:

Photograph of the book   Pencils use creating
                                this book.

EDWARD RUSCHA
(ED-WERD REW-SHAY)
YOUNG ARTIST.

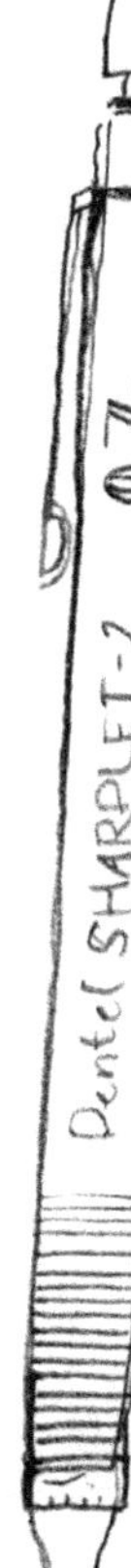

Pentel SHARPLET-2
0.7 AL21